LOOK AROUND AND LISTEN

by Joy Troth Friedman

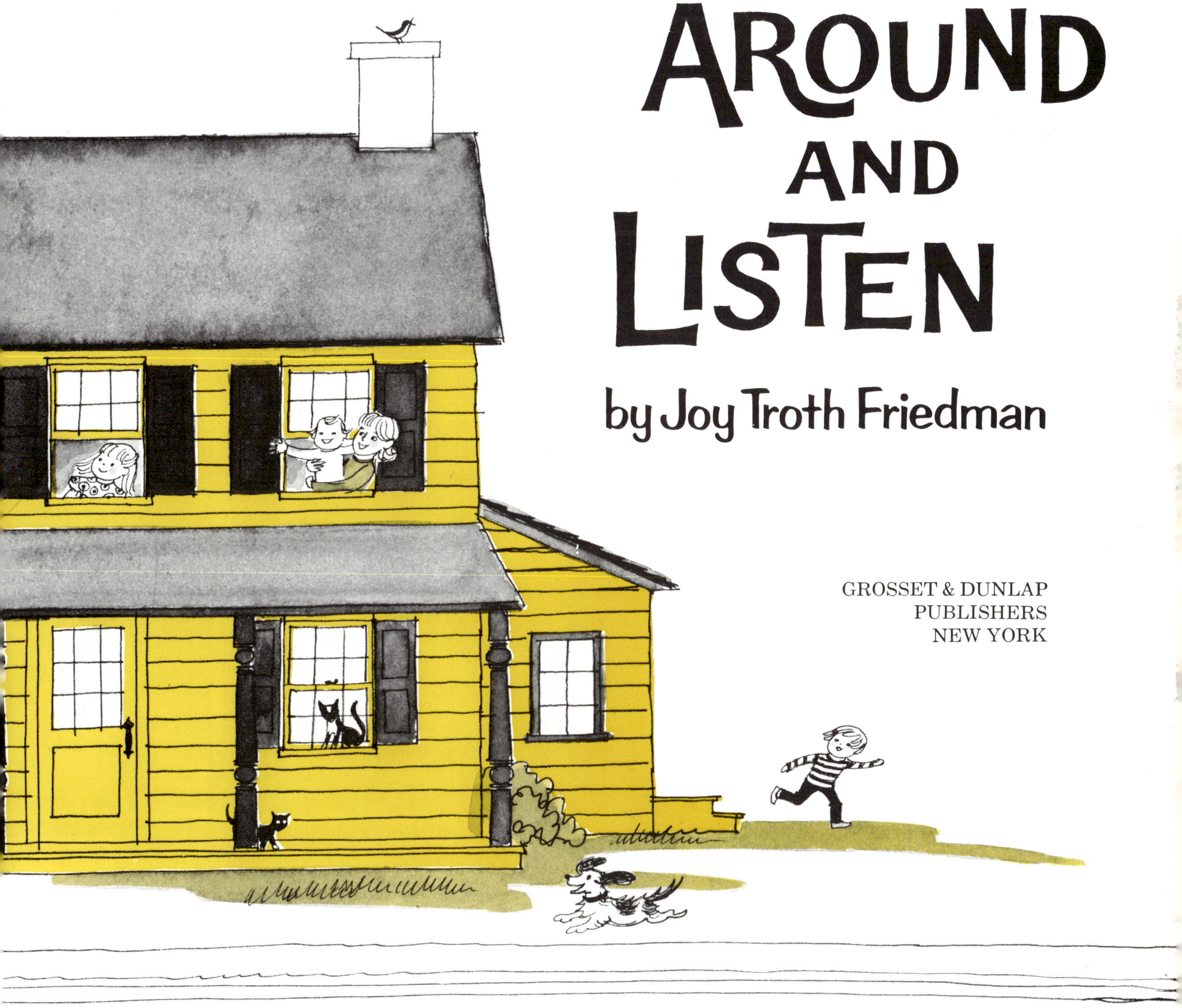

GROSSET & DUNLAP
PUBLISHERS
NEW YORK

Library of Congress Catalog Card Number: 73-16659
ISBN: 0-448-11530-1 (Trade Edition)
ISBN: 0-448-03142-6 (Library Edition)

Published simultaneously in Canada.
Printed in the United States of America.

Sounds

Peckity-peck. The birds are eating seeds.
Whirrrrrrrr! goes the egg beater.
Spit-spat! Bacon is sizzling.
Whew-whew! the tea kettle is whistling!
Popovers are popping in the oven.
Slurp
Drip!
Lappity-lap! The kitten is drinking some milk.
The puppy is drinking water.

Around the House

Betsy is picking up crumbs
with the vacuum cleaner.

Ding-a-ling!
The doorbell is ringing.
Who is there?

Clompety-clomp!
Bill's big boots
make a lot
of noise.

Patty is practicing
her piano lesson.

Tick-
tock!

The metronome
helps her to
keep time.

R-r-r-r-ing!

Swoosh!

Some letters
came through
the mail slot.

Will someone please
answer the telephone?

Do you think anyone can
hear that canary singing?

Time to get up!
The alarm clock
is ringing.

Ding,
ding,
ding,
ding,
ding,
ding!

Sniff, sniff!
Andy
can smell
the bacon
cooking.

Daddy is shaving with his electric razor.

Gurgle, gurgle! The water is running into the bathtub.

The baby is awake. Do you know why she is crying?

The pump in the aquarium is gurgling.

Turtles are VERY quiet.

Squee-e-e-eek! goes the chalk on the blackboard.

Someone forgot to close the door to the music room!

Click! goes the big minute hand on the clock.

Snippity-snip! go the scissors.

Grr-r-r-r-r-r! goes the pencil sharpener.

Sometimes busy places are very noisy, but sometimes they are very quiet!

What do you hear at a Party?

A balloon makes
a funny squeak
when you squeeze it.

Pop! goes the popper.

The kitten is
pretending to be
a scary
Hallowe'en cat.

Swoosh!
A witch is flying by.

What a noisy party!
Do you think everyone is having fun?

Sounds in the Woods in Spring

and in Fall
Hoooo-o-t!
The noisy squirrel family is scurrying about, collecting nuts for the long winter.
If you listen carefully, you might hear an acorn falling.
The woodpecker is putting acorns into the tree trunk.
Tap, tap!
The leaves fall silently . . .
. . . but when you walk through them, they crackle and crunch.
Scratch, scratch! The raccoon is getting his winter bed ready.
Plop!
Do you think fall sounds different from spring?

Sounds on a City Street
yummy Bakery
REAL ESTATE
Screeech!
Screeech!
Screeeeech! go the brakes.
23 Street
The subway train rumbles as it rushes along underground.

Fancy Foot Shoe Shop
Whoooo-Whoooo!
goes the siren
on the patrol car
as it hurries
to answer an
emergency call.
Tweeeeeeet!
goes the
policeman's
whistle.
The policeman
signals the traffic to stop.
Vrrrrrrooom!
Vrrrrrroooom!

Poof!
A-choo!
Flitter!
Tra-la-la!
Voice Piano
DELICATESSEN
jeweler
Paper!
Get your paper!
Woof!
Woof!
Rat-a-tat-tat!
Dum-dum-
de-dum!

Flutter!
Flap, flap!
Bang, bang, bang!
Bong! Bong!
Coooo! Coooo!
BOOKS
Ding-a-ling!
PIPING HOT
Pretty pink posies!
Plinkety-plink!

Too many loud, unpleasant sounds make a lot of NOISE!

Rackety-rackety-rack! goes the jackhammer.

Most city streets are VERY noisy!

what do you hear
Sc-r-eech
goes the spring
on the scale.
Chop, chop!
The butcher is cutting
meat with his cleaver.
Bump,
bump,
bump!
Oops!
There go
the apples.
Plunk!
Squeak!
Arf!

The cat is sunning herself in the window
and taking a cat nap.
Do you think she hears anything?

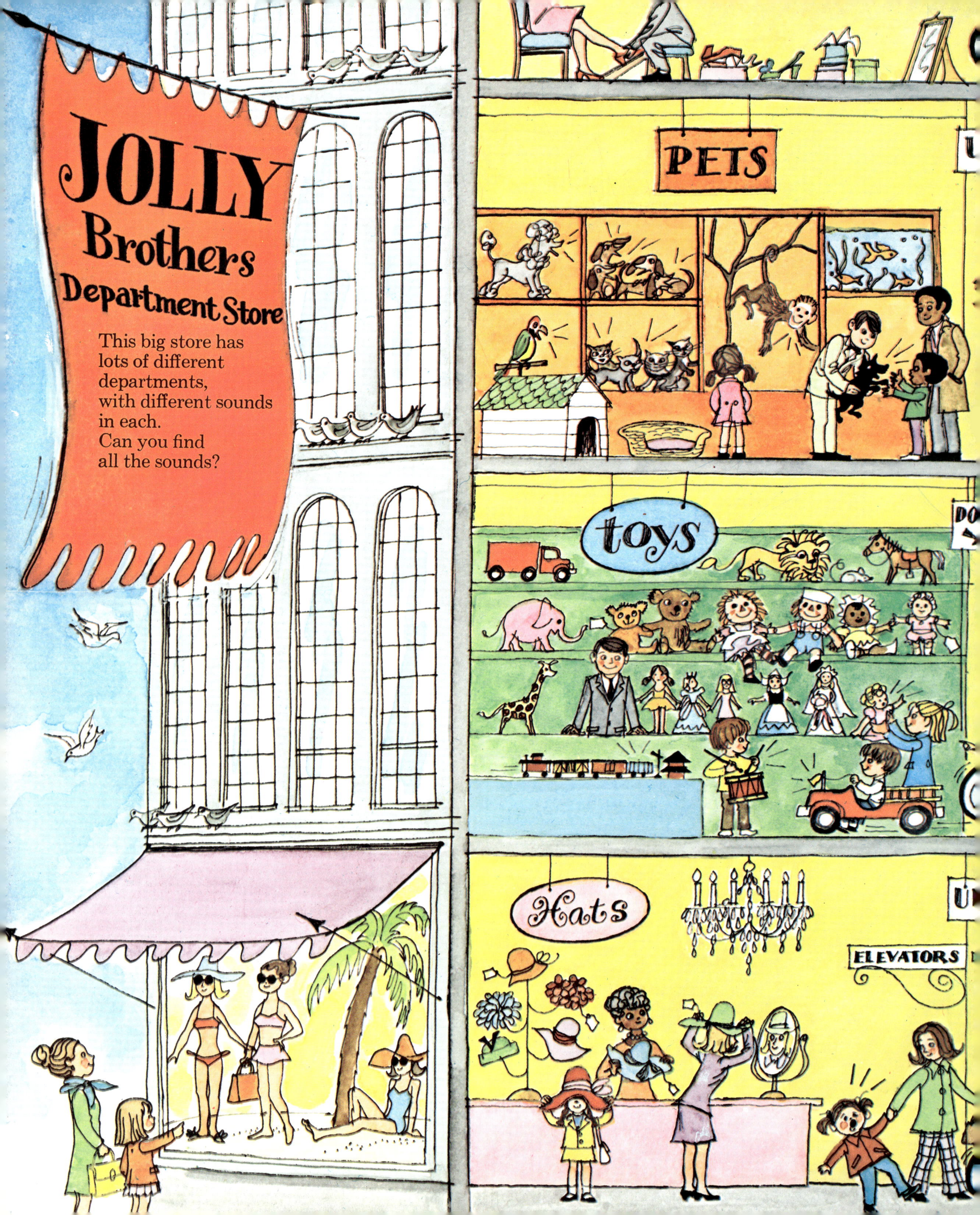
JOLLY
Brothers
Department Store
This big store has
lots of different
departments,
with different sounds
in each.
Can you find
all the sounds?
PETS
toys
Hats
ELEVATORS

SALE
DOWN
MUSIC
UP
APPLIANCES
DOWN
BOOKS

SOUNDS AT THE PLAYGROUND

Creak!
Creak!
Back and forth
go the swings.
Bop,
bop,
bop . . .
goes the
bouncing
ball.
Wow!
Wheeeee!
Sque-e-e-e-e-eak!
goes the merry-go-round.
Hold on tightly!
Hello-o-o-o-o-o!

ZOO SOUNDS
Arf, arf! barks the gnu.
His voice is even stranger than his name.
Hee, hee, hee!
Cheep, cheep!
The playful
monkeys
are chattering
to one another.
Hello,
there!

Tweet!
Squawk!
Caw!
The giraffe seldom speaks, but when he does, his voice is very tiny.
Moooooo
Phew!
The bad-tempered camel huffs and puffs.
Qua-ha! Qua-ha! says the zebra.
Grrrrawrrr!
The tigers and lions have a mighty R-R-ROAR!
The zoo keeper carries the keys for all the animal cages.
Snor-r-rt!
The rhinoceros loves green vegetables, especially peas!
Jingle, jangle!

When the elephant makes a noise, it is called trumpeting. Can you guess why?

The baby elephant loves peanuts. His trunk is like a long finger that puts the peanut in his mouth.

People take lots of pictures at the zoo.

When bears grunt and growl, they sound very fierce.

Penguins seldom talk,
but when they do, they squawk!

bark the playful
sea lions.

The hippopotamus is
smiling for the photographer!

Thump,
thump!

The station agent stamps the date on the ticket.

"Tick-tock, tick-tock," says the old clock. "It's time for the train to come."

Sounds in a Train Station

You can hear the train coming
'way down the track.
It makes a loud, rumbling noise
when it pulls into the station.

Screech! go the brakes
as the train stops
beside the platform.

"All aboard!"
shouts the conductor.

"Wait for me!"

"Goodbye!"

When the train
has gone,
the station
is very quiet. The only sound
is that of the old clock, ticking away until
it's time for the next train to come,

and on the Train

Whoo-o-o-o-o-oo!
Whoo-o-o-o-o-oo!
The engineer is blowing his horn.
Click, click!
The conductor is punching the tickets.
Chitter-chatter
The wheels are spinning along the track!

SOUNDS AT T

airy and scary
snappy and crackly
clippety cloppety
happy and clappy
clangy and bangy
squeaky and freaky

IE CIRCUS are -

Sounds in a Storm

The lightning crackles
and the thunder
rumbles and rolls.

Cars swish through
the puddles as
people hurry home.

Everyone likes to be in a
safe dry place during a storm.
Don't you?

Sounds at the Seashore

Ha, ha, ha, ha!
Kee, kee, kee!
The gulls are laughing and screeching.

Yo-ho!

A sailor is singing a sea chantey.

Ka-room!
The waves are breaking on the shore.

Swish!
The wind is whistling through the tall grass.

Flit-
flit!

A kite is fluttering
in the wind.

Peep, peep, peep!
go the little sandpipers,
scampering back and forth
with the tide.

Scritch,
scratch!

A tiny sandflea is scurrying
into its hole backward.

Bong!
Bong!
goes the bell buoy, rocking to and fro on the waves.
Putt-put
Cr-r-r-ack!
A gull has dropped a clam shell on the rock to break it open.
Swoosh!
The water gurgles and swirls between the rocks.
Crunch! Scrunch!
go the shells and pebbles as they tumble around in the surf.

OOOOooooh-whoooOOOO!

The foghorn is playing a gloomy tune
while the fog creeps in quietly.
It makes
no sound
at
all.

CHUG-A-CHUG-CHUG! A big fishing boat is coming
home with a load of fish.

goes a little motorboat.

When you hold a whelk shell
to your ear, you can hear
the sound of
another ocean
far,
far
away.

pssss-pssss-pssss

If you listen very closely,
you can hear the barnacles
whispering to one another.

Ta-dah
Ta-dah!
The band is playing
a lively tune.
Ooom-pah-pah!
goes the
sousaphone.
Rat-a-tat-tat!
go the
snare drums.
Boop-a-doop!
go the trumpets
and cornets.
Ring-
a-ding-ding!
goes
the
bell-lyr
Boom!
Boom!
Boom!
goes the big bass drum!
Tap-tap go the marching feet.
Tootle-ootle!
go the clarinet
and saxophone.
Clap,
clap!

WHAT DO YOU HEAR AT A
PARADE
Tweeeeeeet!
goes the drum major's whistle.
Whippety-snap! The flags are fluttering in the breeze.
Woof, woof!
Clip-clop! The horses are in step with the music.
Hooray!
Yea!

Sounds In a Restaurant

Whoops!
Chip, chop!
The salad cook
is chopping up
lettuce and onions.
Thumpity-thump!
The pastry cook
is kneading
the dough.
Zzzp!
The chef is
tasting the
soup du jour.
Whack!
The butcher
is slicing some
juicy steaks.
Meow!
A cat tiptoed in
through the back door.
Do you think kitty
will get something to eat?

WHAT DO YOU HEAR IN A MUSEUM?

Museums are very quiet places. Sometimes, all you may hear is the sound of footsteps, echoing through the big rooms.

There is so much to see in a museum,

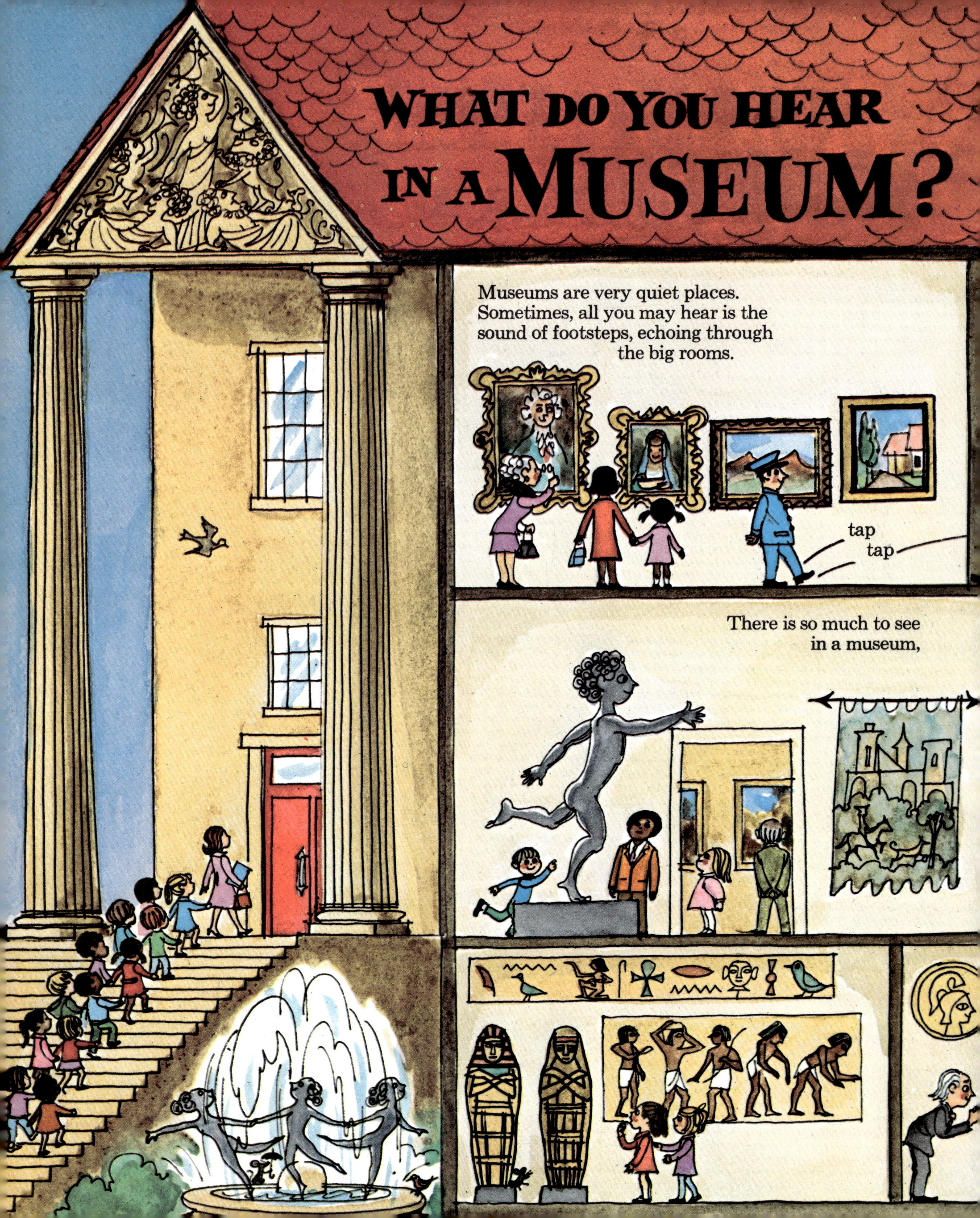

Sometimes you can put on special earphones, and a voice will tell you about what you are seeing.
tap tap
tap tap
that people use their eyes more than their ears.
A guide can tell you about many different things.
Museums are more for looking than for listening!

LISTENING TO

Each instrument has a different sound, but all together they make the sound of a symphony orchestra.

AN ORCHESTRA

The audience is very quiet while the orchestra is playing. What sound do you hear when the music stops?

SOUNDS AT AN AMUSEMENT

PARK

Wow!
Wheeeeee!
Zooooom!
Whirrrrrr!
Ding!
Pop!
Pop!
Bop!
Bop!
Bop!
Yum!
Yum!
Pow!
Wow!

The merriest sound of all is the . . .

. . . . sound of the Merry-Go-Round

What do you hear on a Farm?

ding
dong
Cock-a-doodle
doo!
Heeeeeeer Boss!
oink
cluck
snort!
mooooooooo
moooooo
mooooooo
chug
chug

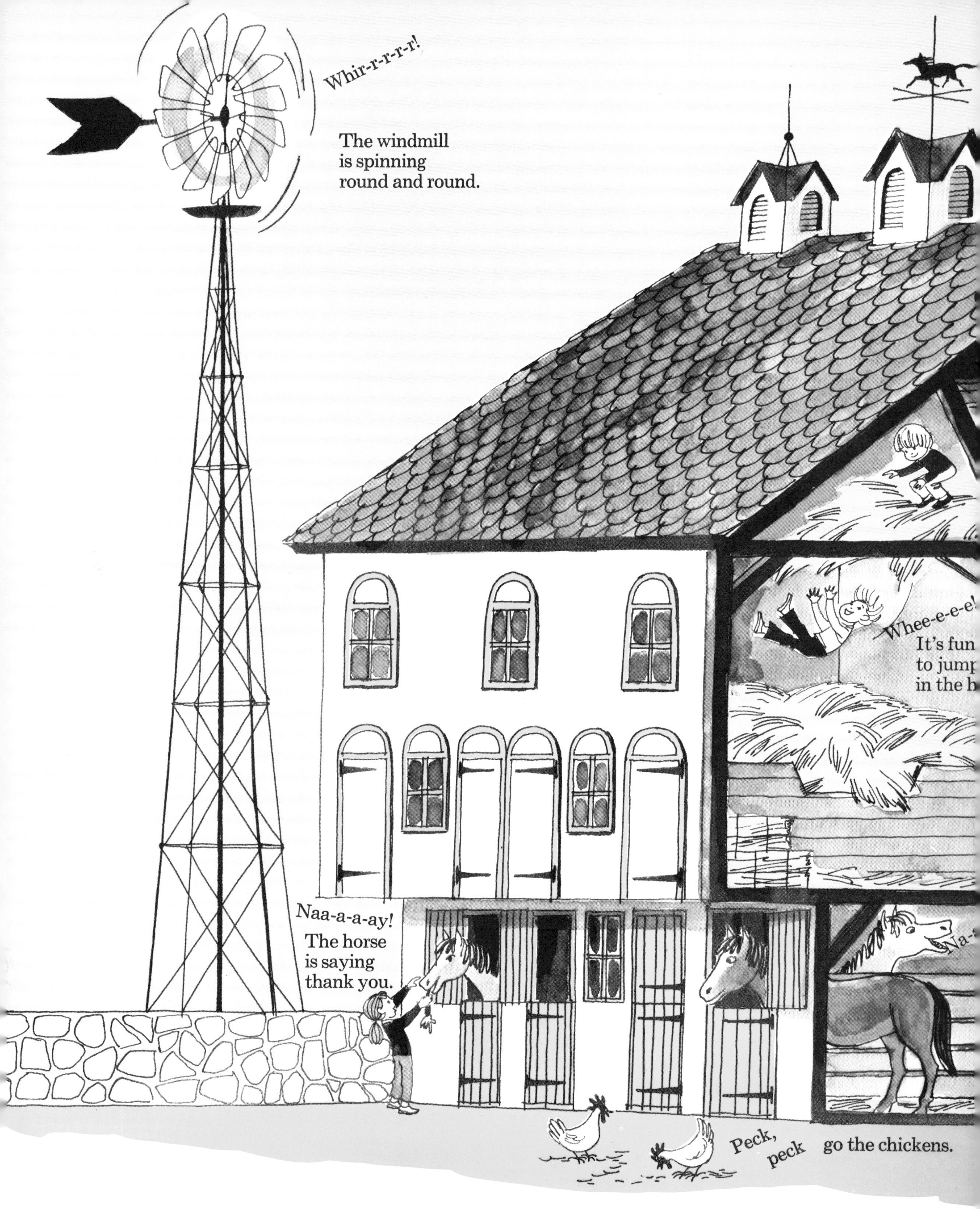

Whir-r-r-r-r!

The windmill
is spinning
round and round.

Whee-e-e-e!

It's fun
to jump
in the h

Naa-a-a-ay!
The horse
is saying
thank you.

Na-

Peck,
peck go the chickens.

The wasps are building a nest in the rafters.

Coo-o-o-o! Coo-o-o-o! go the doves.

The barn swallow is hurrying to feed her hungry babies.

The farmer is milking the cows.

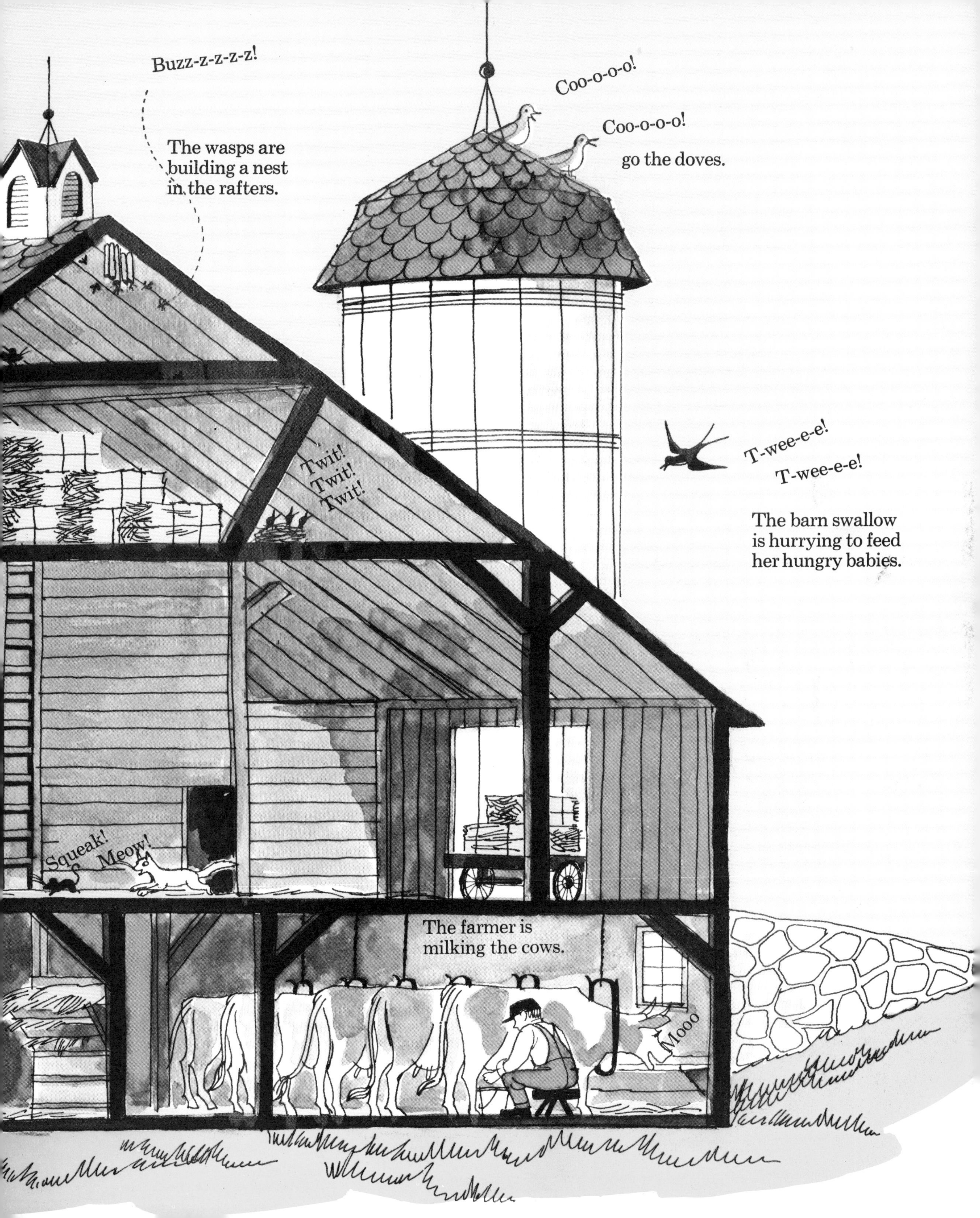

The farmer is storing corn in the corn crib to feed the animals during the winter.

The noisy pigs are pushing and shoving to get their dinner.

The geese like to follow Jimmy when he carries feed to the chickens.

Keeeer-keeeer!
cry the hawks, as they
circle lazily overhead.
Gwoik-
gwoik!
go the hen
turkeys.
Gobble-
gobble!
goes the
big tom
turkey.
Honk!
Honk!
Cock-a-doodle-doo!
crows the
proud
rooster.
Cackle, cackle, cluck!
say the plump
mother hens.

what do you hear in a factory?

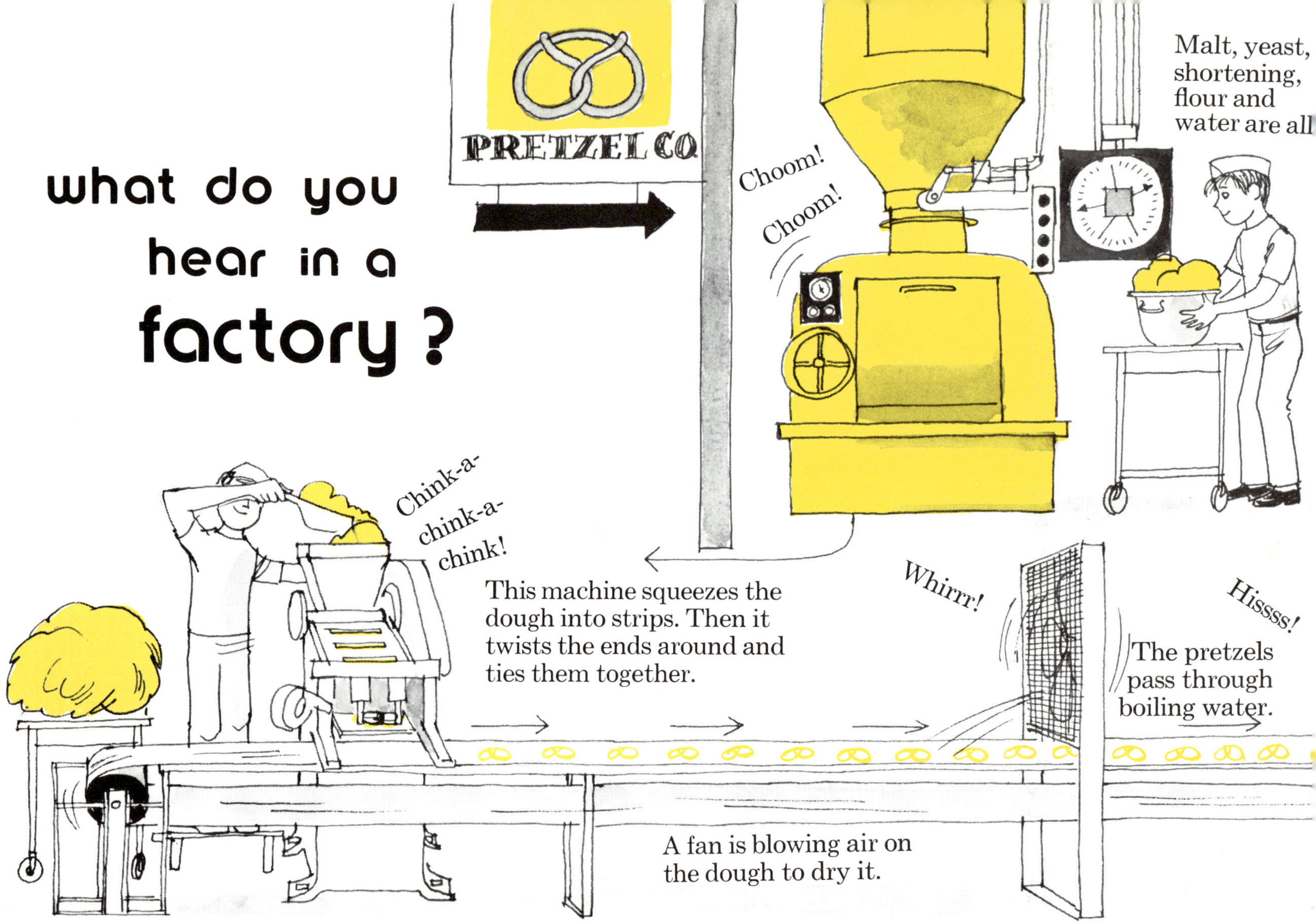

Malt, yeast, shortening, flour and water are all

This machine squeezes the dough into strips. Then it twists the ends around and ties them together.

The pretzels pass through boiling water.

A fan is blowing air on the dough to dry it.

They pass through a dryer to make them crisp.

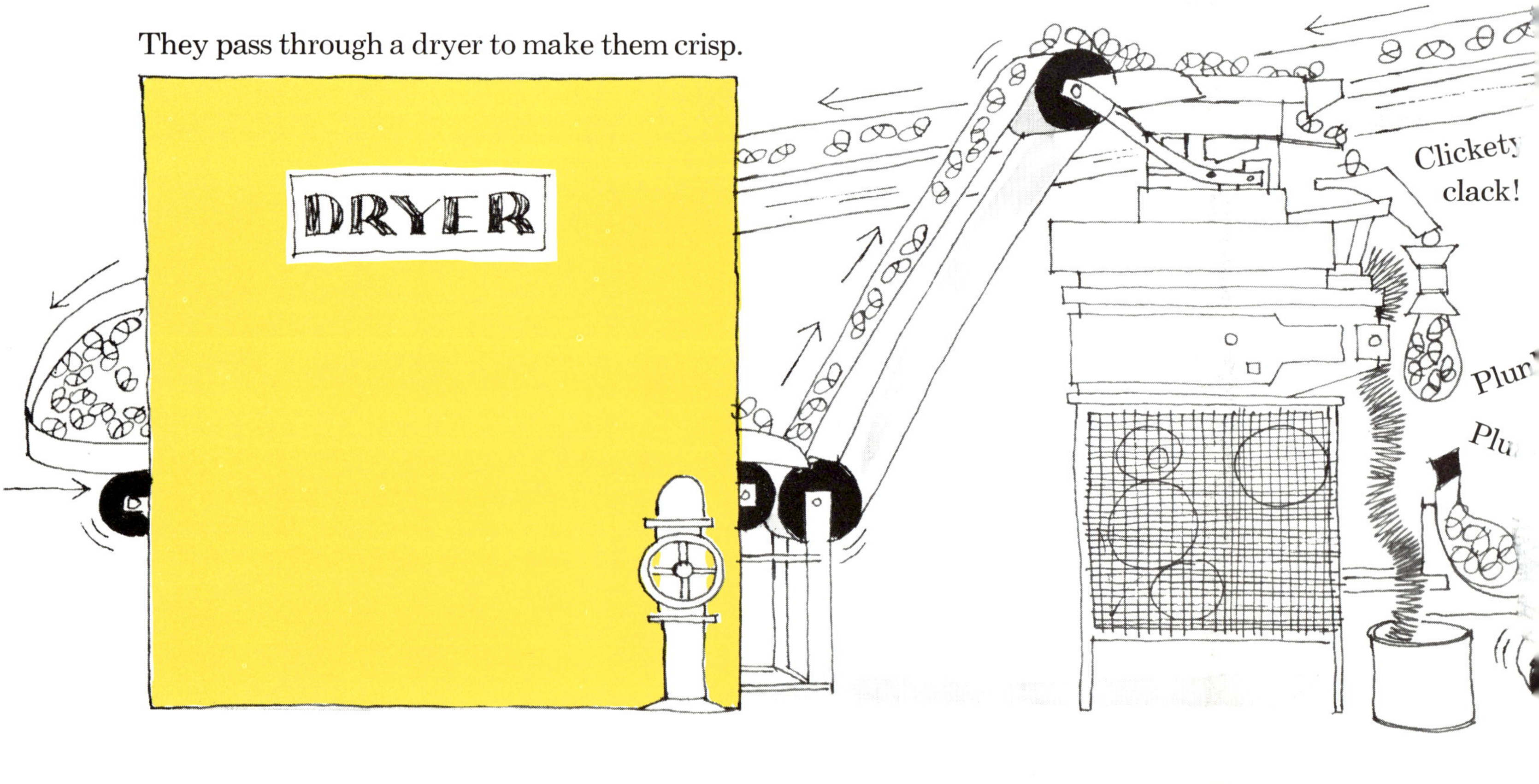

put into a big mixer, which mixes the dough.

On the way to the oven, they are sprinkled with salt.

The big oven is very, very hot. When the pretzels come out, they are golden brown.

The boxes are loaded into trucks and delivered to stores where people may buy them to eat.

The packaging machine drops just the right number of pretzels into each bag and seals it tightly.

The bags are put into boxes.

Do you like to eat pretzels?

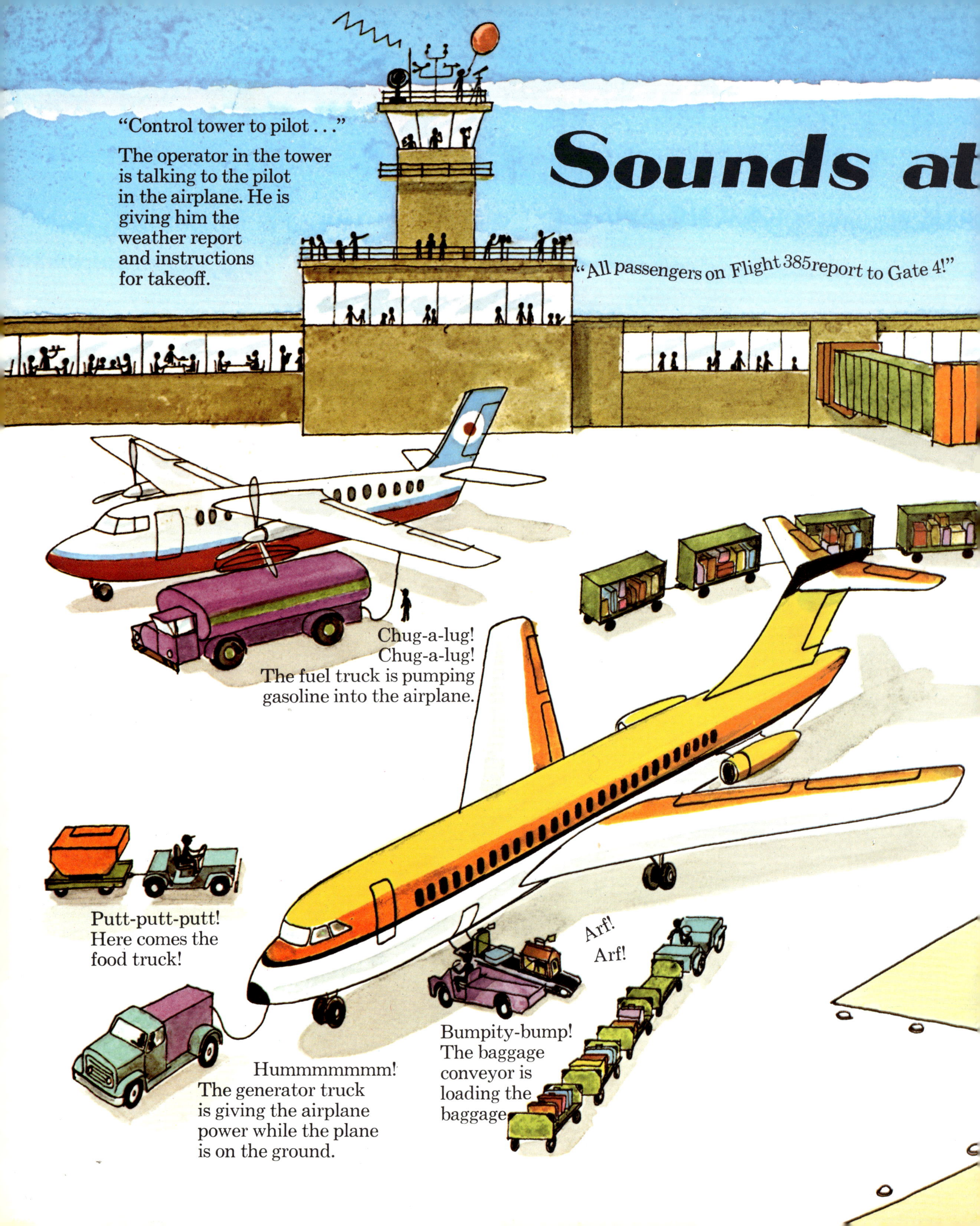

Sounds at

"Control tower to pilot . . ."

The operator in the tower is talking to the pilot in the airplane. He is giving him the weather report and instructions for takeoff.

"All passengers on Flight 385report to Gate 4!"

Chug-a-lug! Chug-a-lug! The fuel truck is pumping gasoline into the airplane.

Putt-putt-putt! Here comes the food truck!

Arf! Arf!

Hummmmmmmm! The generator truck is giving the airplane power while the plane is on the ground.

Bumpity-bump! The baggage conveyor is loading the baggage.

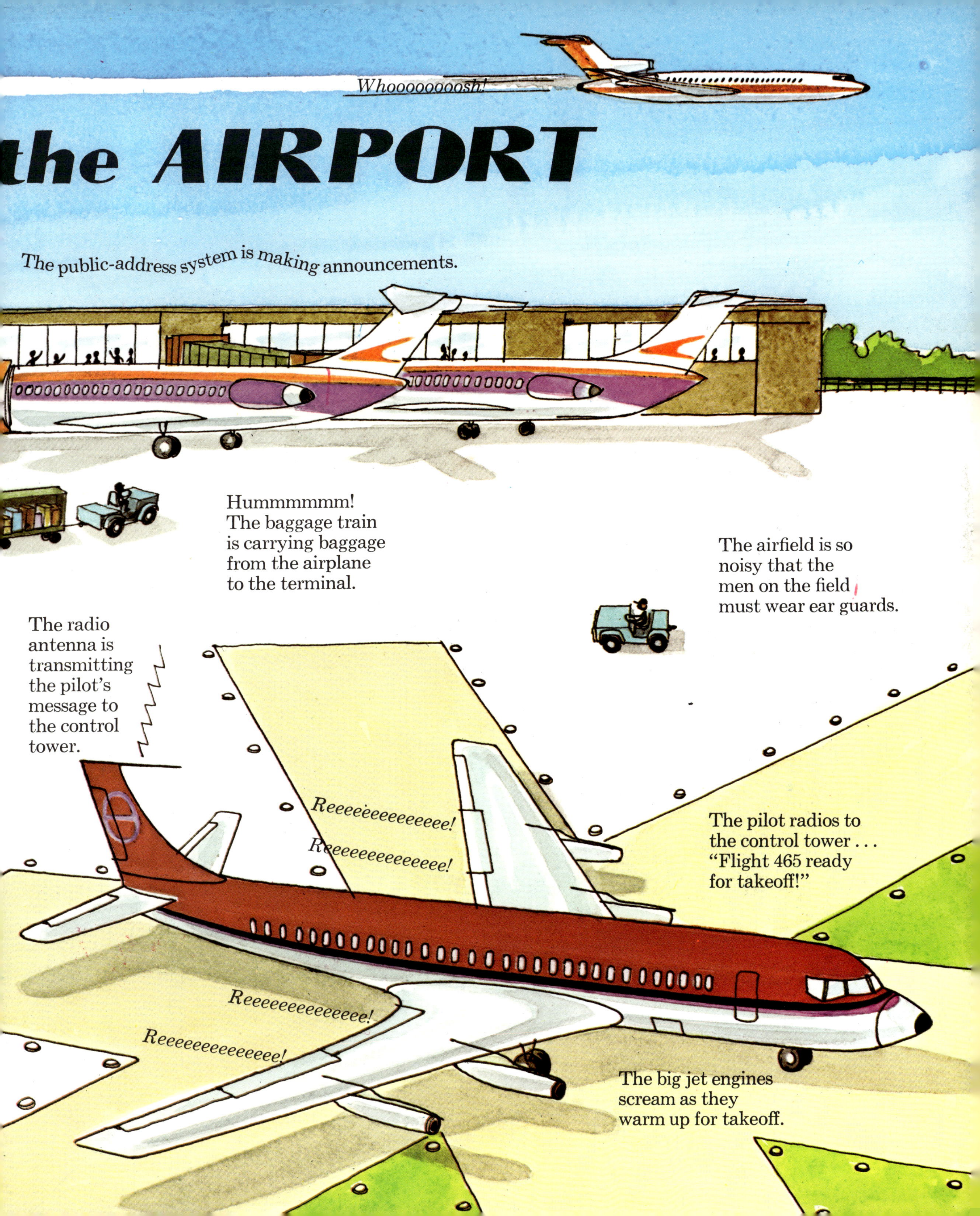
Whoooooooooosh!
the AIRPORT
The public-address system is making announcements.
Hummmmmmm! The baggage train is carrying baggage from the airplane to the terminal.
The airfield is so noisy that the men on the field must wear ear guards.
The radio antenna is transmitting the pilot's message to the control tower.
Reeeeeeeeeeeeeee!
Reeeeeeeeeeeeeee!
The pilot radios to the control tower . . . "Flight 465 ready for takeoff!"
Reeeeeeeeeeeeeee!
Reeeeeeeeeeeeeee!
The big jet engines scream as they warm up for takeoff.

SOUNDS ON

Cree-e-e-eak!
goes the mill wheel
as it turns round
and round.
Quack!
Quack!
Swoosh!
The water pours
over the dam.
Whee-e-e!
Splish!
Splash!
Come on in!
The water's fine!
Cr-r-r-oak!

A River in the Country

Rickety, rackity,
bumpity, bump!
A car is crossing
the old wooden
bridge.

A canoe glides quietly
through the water.
It makes only a tiny swish.

Mother duck is
teaching her
ducklings to swim.

Tommy is sitting
very still, and
not making any
sound at all.

Do you think
he will catch
a fish?

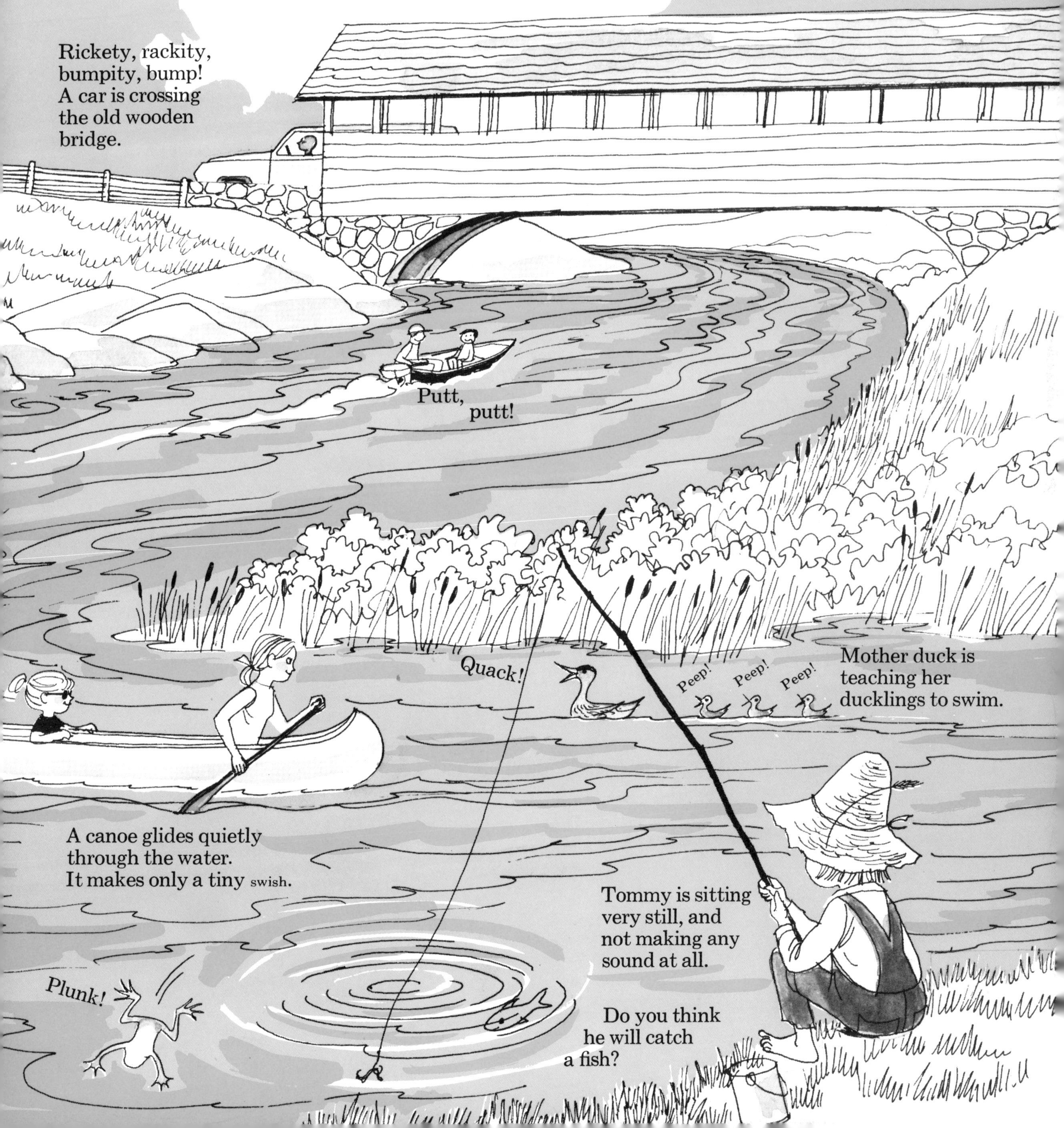

Sounds on a River in the City

Chug, chug, chug!
The ferryboat's
engine works hard
to turn big
propellers.

Toot, toot!
calls the tugboat.

Putt, putt, putt!

Beeeeeep, beeeeeep!
The ferryboat's shrill whistle
answers the tugboat.

Chop, chop, chop!

A helicopter
circles overhead,
watching the
boats come
and go.

Boooooop! Boooooop!

A big ocean liner
sails down the
river to the sea.

Swoosh!
The ferryboat's
propellers
churn the water
into white foam.

Gong! gong!
goes the bell buoy.

Night Sounds in the City

The clock is striking four o'clock.

The street is getting a bath.

This funny bird sings all night long.

The katydids join in a noisy chorus.

A raccoon scrambles up a tree. What does he see?

goes the cheerful cricket.

Night Sounds in the Country

The silent fireflies flit about, blinking their tiny flashlights.

The dog is howling at the full moon.

Squeak! The field mouse scurries home.

The sly fox creeps quietly by on tiptoe.

Night Sounds Around the House

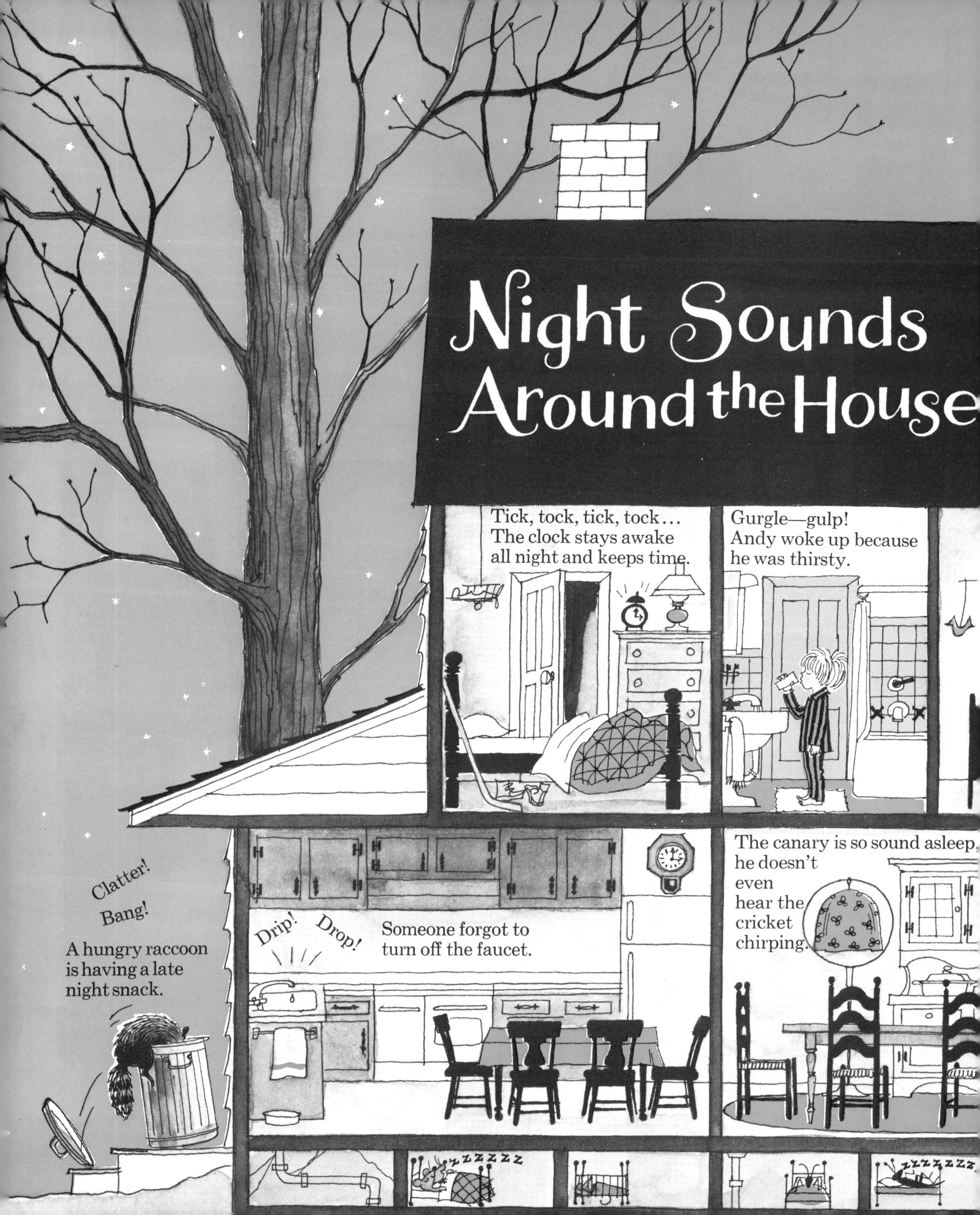

Creak!
go the rafters
when the wind blows.
The cat likes to snuggle in
Betsy's bed.
purrrrrrrrrrr
Daddy makes a lot of noise
when he is sleeping.
The puppy
snores, too.
Bong, bong, bong!
It's three o'clock.
Who
can be
playing
the piano
at
this
hour?
Go back to bed, mouse!
Sometimes
the house is so
quiet at night
that you
can hear
a
pin
drop.
plink!
Chirp!
Shhhhhhhhh!
Sleep tight!